A Member of the International Code Family®

INTERNATIONAL CODE COUNCIL®

INTERNATIONAL

PROPERTY

MAINTENANCE

CODE®

2006

2006 International Property Maintenance Code®

First Printing: January 2006
Second Printing: November 2006
Third Printing: March 2007

ISBN-13: 978-1-58001-263-8 (soft)
ISBN-10: 1-58001-263-9 (soft)
ISBN-13: 978-158001-311-6 (e-document)
ISBN-10: 1-58001-311-2 (e-document)

PRINTED IN THE U.S.A.

PREFACE

Introduction

Internationally, code officials recognize the need for a modern, up-to-date property maintenance code governing the maintenance of existing buildings. The *International Property Maintenance Code®*, in this 2006 edition, is designed to meet this need through model code regulations that contain clear and specific property maintenance requirements with required property improvement provisions.

This 2006 edition is fully compatible with all *International Codes®* (I-Codes®) published by the International Code Council (ICC)®, including the *International Building Code®*, ICC *Electrical Code®—Administrative Provisions*, *International Energy Conservation Code®*, *International Existing Building Code®*, *International Fire Code®*, *International Fuel Gas Code®*, *International Mechanical Code®*, ICC *Performance Code®*, *International Plumbing Code®*, *International Private Sewage Disposal Code®*, *International Residential Code®*, *International Wildland-Urban Interface Code™* and *International Zoning Code®*.

The *International Property Maintenance Code* provisions provide many benefits, among which is the model code development process that offers an international forum for code officials and other interested parties to discuss performance and prescriptive code requirements. This forum provides an excellent arena to debate proposed revisions. This model code also encourages international consistency in the application of provisions.

Development

The first edition of the *International Property Maintenance Code* (1998) was the culmination of an effort initiated in 1996 by a code development committee appointed by ICC and consisting of representatives of the three statutory members of the International Code Council at that time, including: Building Officials and Code Administrators International, Inc. (BOCA), International Conference of Building Officials (ICBO) and Southern Building Code Congress International (SBCCI). The committee drafted a comprehensive set of regulations for existing buildings that was consistent with the existing model property maintenance codes at the time. This 2006 edition presents the code as originally issued, with changes reflected through the previous 2003 editions and further changes developed through the ICC Code Development Process through 2005. A new edition of the code is promulgated every three years.

This code is founded on principles intended to establish provisions consistent with the scope of a property maintenance code that adequately protects public health, safety and welfare; provisions that do not unnecessarily increase construction costs; provisions that do not restrict the use of new materials, products or methods of construction; and provisions that do not give preferential treatment to particular types or classes of materials, products or methods of construction.

Adoption

The *International Property Maintenance Code* is available for adoption and use by jurisdictions internationally. Its use within a governmental jurisdiction is intended to be accomplished through adoption by reference in accordance with proceedings establishing the jurisdiction's laws. At the time of adoption, jurisdictions should insert the appropriate information in provisions requiring specific local information, such as the name of the adopting jurisdiction. These locations are shown in bracketed words in small capital letters in the code and in the sample ordinance. The sample adoption ordinance on page v addresses several key elements of a code adoption ordinance, including the information required for insertion into the code text.

Maintenance

The *International Property Maintenance Code* is kept up to date through the review of proposed changes submitted by code enforcing officials, industry representatives, design professionals and other interested parties. Proposed changes are carefully considered through an open code development process in which all interested and affected parties may participate.

The contents of this work are subject to change both through the Code Development Cycles and the governmental body that enacts the code into law. For more information regarding the code development process, contact the Codes and Standards Development Department of the International Code Council.

While the development procedure of the *International Property Maintenance Code* ensures the highest degree of care, ICC, its membership and those participating in the development of this code do not accept any liability resulting from compliance or noncompliance with the provisions because ICC does not have the power or authority to police or enforce compliance with the contents of this code. Only the governmental body that enacts the code into law has such authority.

Letter Designations in Front of Section Numbers

In each code development cycle, proposed changes to this code are considered at the Code Development Hearings by the ICC Property Maintenance/Zoning Code Development Committee, whose action constitutes a recommendation to the voting membership for final action on the proposed changes. Proposed changes to a code section having a number beginning with a letter in brackets are considered by a different code development committee. For example, proposed changes to code sections that have the letter [F] in front of them (e.g., [F] 704.1) are considered by the International Fire Code Development Committee at the Code Development Hearings.

The content of sections in this code that begin with a letter designation are maintained by another code development committee in accordance with the following:

[F] = International Fire Code Development Committee;

[P] = International Plumbing Code Development Committee;

[F] = International Fire Code Development Committee; and

[B] = International Building Code Development Committee.

Marginal Markings

Solid vertical lines in the margins within the body of the code indicating a technical change from the requirements of the previous edition. Deletion indicators in the form of an arrow (➡) are provided in the margin where an entire section, paragraph, exception or table has been deleted or an item in a list of items or a table has been deleted.

ORDINANCE

The *International Codes* are designed and promulgated to be adopted by reference by ordinance. Jurisdictions wishing to adopt the 2006 *International Property Maintenance Code* as an enforceable regulation governing existing structures and premises should ensure that certain factual information is included in the adopting ordinance at the time adoption is being considered by the appropriate governmental body. The following sample adoption ordinance addresses several key elements of a code adoption ordinance, including the information required for insertion into the code text.

SAMPLE ORDINANCE FOR ADOPTION OF
THE *INTERNATIONAL PROPERTY MAINTENANCE CODE*
ORDINANCE NO._____

An ordinance of the **[JURISDICTION]** adopting the 2006 edition of the *International Property Maintenance Code*, regulating and governing the conditions and maintenance of all property, buildings and structures; by providing the standards for supplied utilities and facilities and other physical things and conditions essential to ensure that structures are safe, sanitary and fit for occupation and use; and the condemnation of buildings and structures unfit for human occupancy and use, and the demolition of such existing structures in the **[JURISDICTION]**; providing for the issuance of permits and collection of fees therefor; repealing Ordinance No. _____ of the **[JURISDICTION]** and all other ordinances and parts of the ordinances in conflict therewith.

The **[GOVERNING BODY]** of the **[JURISDICTION]** does ordain as follows:

Section 1. That a certain document, three (3) copies of which are on file in the office of the **[TITLE OF JURISDICTION'S KEEPER OF RECORDS]** of **[NAME OF JURISDICTION]**, being marked and designated as the *International Property Maintenance Code*, 2006 edition, as published by the International Code Council, be and is hereby adopted as the Property Maintenance Code of the **[JURISDICTION]**, in the State of **[STATE NAME]** for regulating and governing the conditions and maintenance of all property, buildings and structures; by providing the standards for supplied utilities and facilities and other physical things and conditions essential to ensure that structures are safe, sanitary and fit for occupation and use; and the condemnation of buildings and structures unfit for human occupancy and use, and the demolition of such existing structures as herein provided; providing for the issuance of permits and collection of fees therefor; and each and all of the regulations, provisions, penalties, conditions and terms of said Property Maintenance Code on file in the office of the **[JURISDICTION]** are hereby referred to, adopted, and made a part hereof, as if fully set out in this ordinance, with the additions, insertions, deletions and changes, if any, prescribed in Section 2 of this ordinance.

Section 2. The following sections are hereby revised:

Section 101.1. Insert: **[NAME OF JURISDICTION]**

Section 103.5. Insert: **[APPROPRIATE SCHEDULE]**

Section 302.4. Insert: **[HEIGHT IN INCHES]**

Section 304.14. Insert: **[DATES IN TWO LOCATIONS]**

Section 602.3. Insert: **[DATES IN TWO LOCATIONS]**

Section 602.4. Insert: **[DATES IN TWO LOCATIONS]**

Section 3. That Ordinance No. _____ of **[JURISDICTION]** entitled **[FILL IN HERE THE COMPLETE TITLE OF THE ORDINANCE OR ORDINANCES IN EFFECT AT THE PRESENT TIME SO THAT THEY WILL BE REPEALED BY DEFINITE MENTION]** and all other ordinances or parts of ordinances in conflict herewith are hereby repealed.

Section 4. That if any section, subsection, sentence, clause or phrase of this ordinance is, for any reason, held to be unconstitutional, such decision shall not affect the validity of the remaining portions of this ordinance. The **[GOVERNING BODY]** hereby declares that it would have passed this ordinance, and each section, subsection, clause or phrase thereof, irrespective of the fact that any one or more sections, subsections, sentences, clauses and phrases be declared unconstitutional.

Section 5. That nothing in this ordinance or in the Property Maintenance Code hereby adopted shall be construed to affect any suit or proceeding impending in any court, or any rights acquired, or liability incurred, or any cause or causes of action acquired or existing, under any act or ordinance hereby repealed as cited in Section 3 of this ordinance; nor shall any just or legal right or remedy of any character be lost, impaired or affected by this ordinance.

Section 6. That the **[JURISDICTION'S KEEPER OF RECORDS]** is hereby ordered and directed to cause this ordinance to be published. (An additional provision may be required to direct the number of times the ordinance is to be published and to specify that it is to be in a newspaper in general circulation. Posting may also be required.)

Section 7. That this ordinance and the rules, regulations, provisions, requirements, orders and matters established and adopted hereby shall take effect and be in full force and effect **[TIME PERIOD]** from and after the date of its final passage and adoption.

TABLE OF CONTENTS

CHAPTER 1

ADMINISTRATION

SECTION 101
GENERAL

101.1 Title. These regulations shall be known as the *Property Maintenance Code* of [NAME OF JURISDICTION], hereinafter referred to as "this code."

101.2 Scope. The provisions of this code shall apply to all existing residential and nonresidential structures and all existing premises and constitute minimum requirements and standards for premises, structures, equipment and facilities for light, ventilation, space, heating, sanitation, protection from the elements, life safety, safety from fire and other hazards, and for safe and sanitary maintenance; the responsibility of owners, operators and occupants; the occupancy of existing structures and premises, and for administration, enforcement and penalties.

101.3 Intent. This code shall be construed to secure its expressed intent, which is to ensure public health, safety and welfare in so far as they are affected by the continued occupancy and maintenance of structures and premises. Existing structures and premises that do not comply with these provisions shall be altered or repaired to provide a minimum level of health and safety as required herein.

101.4 Severability. If a section, subsection, sentence, clause or phrase of this code is, for any reason, held to be unconstitutional, such decision shall not affect the validity of the remaining portions of this code.

SECTION 102
APPLICABILITY

102.1 General. The provisions of this code shall apply to all matters affecting or relating to structures and premises, as set forth in Section 101. Where, in a specific case, different sections of this code specify different requirements, the most restrictive shall govern.

102.2 Maintenance. Equipment, systems, devices and safeguards required by this code or a previous regulation or code under which the structure or premises was constructed, altered or repaired shall be maintained in good working order. No owner, operator or occupant shall cause any service, facility, equipment or utility which is required under this section to be removed from or shut off from or discontinued for any occupied dwelling, except for such temporary interruption as necessary while repairs or alterations are in progress. The requirements of this code are not intended to provide the basis for removal or abrogation of fire protection and safety systems and devices in existing structures. Except as otherwise specified herein, the owner or the owner's designated agent shall be responsible for the maintenance of buildings, structures and premises.

102.3 Application of other codes. Repairs, additions or alterations to a structure, or changes of occupancy, shall be done in accordance with the procedures and provisions of the *International Building Code, International Fuel Gas Code, International Mechanical Code* and the ICC *Electrical Code*. Nothing in this code shall be construed to cancel, modify or set aside any provision of the *International Zoning Code*.

102.4 Existing remedies. The provisions in this code shall not be construed to abolish or impair existing remedies of the jurisdiction or its officers or agencies relating to the removal or demolition of any structure which is dangerous, unsafe and insanitary.

102.5 Workmanship. Repairs, maintenance work, alterations or installations which are caused directly or indirectly by the enforcement of this code shall be executed and installed in a workmanlike manner and installed in accordance with the manufacturer's installation instructions.

102.6 Historic buildings. The provisions of this code shall not be mandatory for existing buildings or structures designated as historic buildings when such buildings or structures are judged by the code official to be safe and in the public interest of health, safety and welfare.

102.7 Referenced codes and standards. The codes and standards referenced in this code shall be those that are listed in Chapter 8 and considered part of the requirements of this code to the prescribed extent of each such reference. Where differences occur between provisions of this code and the referenced standards, the provisions of this code shall apply.

102.8 Requirements not covered by code. Requirements necessary for the strength, stability or proper operation of an existing fixture, structure or equipment, or for the public safety, health and general welfare, not specifically covered by this code, shall be determined by the code official.

SECTION 103
DEPARTMENT OF PROPERTY
MAINTENANCE INSPECTION

103.1 General. The department of property maintenance inspection is hereby created and the executive official in charge thereof shall be known as the code official.

103.2 Appointment. The code official shall be appointed by the chief appointing authority of the jurisdiction; and the code official shall not be removed from office except for cause and after full opportunity to be heard on specific and relevant charges by and before the appointing authority.

103.3 Deputies. In accordance with the prescribed procedures of this jurisdiction and with the concurrence of the appointing authority, the code official shall have the authority to appoint a deputy code official, other related technical officers, inspectors and other employees.

103.4 Liability. The code official, officer or employee charged with the enforcement of this code, while acting for the jurisdiction, shall not thereby be rendered liable personally, and is hereby relieved from all personal liability for any damage accruing to persons or property as a result of an act required or permitted in the discharge of official duties.

Any suit instituted against any officer or employee because of an act performed by that officer or employee in the lawful discharge of duties and under the provisions of this code shall be defended by the legal representative of the jurisdiction until the final termination of the proceedings. The code official or any subordinate shall not be liable for costs in an action, suit or proceeding that is instituted in pursuance of the provisions of this code; and any officer of the department of property maintenance inspection, acting in good faith and without malice, shall be free from liability for acts performed under any of its provisions or by reason of any act or omission in the performance of official duties in connection therewith.

103.5 Fees. The fees for activities and services performed by the department in carrying out its responsibilities under this code shall be as indicated in the following schedule.

[JURISDICTION TO INSERT APPROPRIATE SCHEDULE.]

SECTION 104
DUTIES AND POWERS OF THE CODE OFFICIAL

104.1 General. The code official shall enforce the provisions of this code.

104.2 Rule-making authority. The code official shall have authority as necessary in the interest of public health, safety and general welfare, to adopt and promulgate rules and procedures; to interpret and implement the provisions of this code; to secure the intent thereof; and to designate requirements applicable because of local climatic or other conditions. Such rules shall not have the effect of waiving structural or fire performance requirements specifically provided for in this code, or of violating accepted engineering methods involving public safety.

104.3 Inspections. The code official shall make all of the required inspections, or shall accept reports of inspection by approved agencies or individuals. All reports of such inspections shall be in writing and be certified by a responsible officer of such approved agency or by the responsible individual. The code official is authorized to engage such expert opinion as deemed necessary to report upon unusual technical issues that arise, subject to the approval of the appointing authority.

104.4 Right of entry. The code official is authorized to enter the structure or premises at reasonable times to inspect subject to constitutional restrictions on unreasonable searches and seizures. If entry is refused or not obtained, the code official is authorized to pursue recourse as provided by law.

104.5 Identification. The code official shall carry proper identification when inspecting structures or premises in the performance of duties under this code.

104.6 Notices and orders. The code official shall issue all necessary notices or orders to ensure compliance with this code.

104.7 Department records. The code official shall keep official records of all business and activities of the department specified in the provisions of this code. Such records shall be retained in the official records as long as the building or structure to which such records relate remains in existence, unless otherwise provided for by other regulations.

SECTION 105
APPROVAL

105.1 Modifications. Whenever there are practical difficulties involved in carrying out the provisions of this code, the code official shall have the authority to grant modifications for individual cases, provided the code official shall first find that special individual reason makes the strict letter of this code impractical and the modification is in compliance with the intent and purpose of this code and that such modification does not lessen health, life and fire safety requirements. The details of action granting modifications shall be recorded and entered in the department files.

105.2 Alternative materials, methods and equipment. The provisions of this code are not intended to prevent the installation of any material or to prohibit any method of construction not specifically prescribed by this code, provided that any such alternative has been approved. An alternative material or method of construction shall be approved where the code official finds that the proposed design is satisfactory and complies with the intent of the provisions of this code, and that the material, method or work offered is, for the purpose intended, at least the equivalent of that prescribed in this code in quality, strength, effectiveness, fire resistance, durability and safety.

105.3 Required testing. Whenever there is insufficient evidence of compliance with the provisions of this code, or evidence that a material or method does not conform to the requirements of this code, or in order to substantiate claims for alternative materials or methods, the code official shall have the authority to require tests to be made as evidence of compliance at no expense to the jurisdiction.

105.3.1 Test methods. Test methods shall be as specified in this code or by other recognized test standards. In the absence of recognized and accepted test methods, the code official shall be permitted to approve appropriate testing procedures performed by an approved agency.

105.3.2 Test reports. Reports of tests shall be retained by the code official for the period required for retention of public records.

105.4 Material and equipment reuse. Materials, equipment and devices shall not be reused unless such elements are in good repair or have been reconditioned and tested when necessary, placed in good and proper working condition and approved.

SECTION 106
VIOLATIONS

106.1 Unlawful acts. It shall be unlawful for a person, firm or corporation to be in conflict with or in violation of any of the provisions of this code.

106.2 Notice of violation. The code official shall serve a notice of violation or order in accordance with Section 107.

106.3 Prosecution of violation. Any person failing to comply with a notice of violation or order served in accordance with Section 107 shall be deemed guilty of a misdemeanor or civil infraction as determined by the local municipality, and the violation shall be deemed a strict liability offense. If the notice of violation is not complied with, the code official shall institute the appropriate proceeding at law or in equity to restrain, correct or abate such violation, or to require the removal or termination of the unlawful occupancy of the structure in violation of the provisions of this code or of the order or direction made pursuant thereto. Any action taken by the authority having jurisdiction on such premises shall be charged against the real estate upon which the structure is located and shall be a lien upon such real estate.

106.4 Violation penalties. Any person who shall violate a provision of this code, or fail to comply therewith, or with any of the requirements thereof, shall be prosecuted within the limits provided by state or local laws. Each day that a violation continues after due notice has been served shall be deemed a separate offense.

106.5 Abatement of violation. The imposition of the penalties herein prescribed shall not preclude the legal officer of the jurisdiction from instituting appropriate action to restrain, correct or abate a violation, or to prevent illegal occupancy of a building, structure or premises, or to stop an illegal act, conduct, business or utilization of the building, structure or premises.

SECTION 107
NOTICES AND ORDERS

107.1 Notice to person responsible. Whenever the code official determines that there has been a violation of this code or has grounds to believe that a violation has occurred, notice shall be given in the manner prescribed in Sections 107.2 and 107.3 to the person responsible for the violation as specified in this code. Notices for condemnation procedures shall also comply with Section 108.3.

107.2 Form. Such notice prescribed in Section 107.1 shall be in accordance with all of the following:

1. Be in writing.
2. Include a description of the real estate sufficient for identification.
3. Include a statement of the violation or violations and why the notice is being issued.
4. Include a correction order allowing a reasonable time to make the repairs and improvements required to bring the dwelling unit or structure into compliance with the provisions of this code.
5. Inform the property owner of the right to appeal.
6. Include a statement of the right to file a lien in accordance with Section 106.3.

107.3 Method of service. Such notice shall be deemed to be properly served if a copy thereof is:

1. Delivered personally;
2. Sent by certified or first-class mail addressed to the last known address; or
3. If the notice is returned showing that the letter was not delivered, a copy thereof shall be posted in a conspicuous place in or about the structure affected by such notice.

107.4 Penalties. Penalties for noncompliance with orders and notices shall be as set forth in Section 106.4.

107.5 Transfer of ownership. It shall be unlawful for the owner of any dwelling unit or structure who has received a compliance order or upon whom a notice of violation has been served to sell, transfer, mortgage, lease or otherwise dispose of such dwelling unit or structure to another until the provisions of the compliance order or notice of violation have been complied with, or until such owner shall first furnish the grantee, transferee, mortgagee or lessee a true copy of any compliance order or notice of violation issued by the code official and shall furnish to the code official a signed and notarized statement from the grantee, transferee, mortgagee or lessee, acknowledging the receipt of such compliance order or notice of violation and fully accepting the responsibility without condition for making the corrections or repairs required by such compliance order or notice of violation.

SECTION 108
UNSAFE STRUCTURES AND EQUIPMENT

108.1 General. When a structure or equipment is found by the code official to be unsafe, or when a structure is found unfit for human occupancy, or is found unlawful, such structure shall be condemned pursuant to the provisions of this code.

108.1.1 Unsafe structures. An unsafe structure is one that is found to be dangerous to the life, health, property or safety of the public or the occupants of the structure by not providing minimum safeguards to protect or warn occupants in the event of fire, or because such structure contains unsafe equipment or is so damaged, decayed, dilapidated, structurally unsafe or of such faulty construction or unstable foundation, that partial or complete collapse is possible.

108.1.2 Unsafe equipment. Unsafe equipment includes any boiler, heating equipment, elevator, moving stairway, electrical wiring or device, flammable liquid containers or other equipment on the premises or within the structure which is in such disrepair or condition that such equipment is a hazard to life, health, property or safety of the public or occupants of the premises or structure.

108.1.3 Structure unfit for human occupancy. A structure is unfit for human occupancy whenever the code official finds that such structure is unsafe, unlawful or, because of the degree to which the structure is in disrepair or lacks maintenance, is insanitary, vermin or rat infested, contains filth and contamination, or lacks ventilation, illumination, sanitary or heating facilities or other essential equipment required by this code, or because the location of the structure constitutes a hazard to the occupants of the structure or to the public.

108.1.4 Unlawful structure. An unlawful structure is one found in whole or in part to be occupied by more persons than permitted under this code, or was erected, altered or occupied contrary to law.

108.2 Closing of vacant structures. If the structure is vacant and unfit for human habitation and occupancy, and is not in danger of structural collapse, the code official is authorized to post a placard of condemnation on the premises and order the structure closed up so as not to be an attractive nuisance. Upon failure of the owner to close up the premises within the time specified in the order, the code official shall cause the premises to be closed and secured through any available public agency or by contract or arrangement by private persons and the cost thereof shall be charged against the real estate upon which the structure is located and shall be a lien upon such real estate and may be collected by any other legal resource.

108.3 Notice. Whenever the code official has condemned a structure or equipment under the provisions of this section, notice shall be posted in a conspicuous place in or about the structure affected by such notice and served on the owner or the person or persons responsible for the structure or equipment in accordance with Section 107.3. If the notice pertains to equipment, it shall also be placed on the condemned equipment. The notice shall be in the form prescribed in Section 107.2.

108.4 Placarding. Upon failure of the owner or person responsible to comply with the notice provisions within the time given, the code official shall post on the premises or on defective equipment a placard bearing the word "Condemned" and a statement of the penalties provided for occupying the premises, operating the equipment or removing the placard.

108.4.1 Placard removal. The code official shall remove the condemnation placard whenever the defect or defects upon which the condemnation and placarding action were based have been eliminated. Any person who defaces or removes a condemnation placard without the approval of the code official shall be subject to the penalties provided by this code.

108.5 Prohibited occupancy. Any occupied structure condemned and placarded by the code official shall be vacated as ordered by the code official. Any person who shall occupy a placarded premises or shall operate placarded equipment, and any owner or any person responsible for the premises who shall let anyone occupy a placarded premises or operate placarded equipment shall be liable for the penalties provided by this code.

SECTION 109
EMERGENCY MEASURES

109.1 Imminent danger. When, in the opinion of the code official, there is imminent danger of failure or collapse of a building or structure which endangers life, or when any structure or part of a structure has fallen and life is endangered by the occupation of the structure, or when there is actual or potential danger to the building occupants or those in the proximity of any structure because of explosives, explosive fumes or vapors or the presence of toxic fumes, gases or materials, or operation of defective or dangerous equipment, the code official is hereby authorized and empowered to order and require the occupants to vacate the premises forthwith. The code official shall cause to be posted at each entrance to such structure a notice reading as follows: "This Structure Is Unsafe and Its Occupancy Has Been Prohibited by the Code Official." It shall be unlawful for any person to enter such structure except for the purpose of securing the structure, making the required repairs, removing the hazardous condition or of demolishing the same.

109.2 Temporary safeguards. Notwithstanding other provisions of this code, whenever, in the opinion of the code official, there is imminent danger due to an unsafe condition, the code official shall order the necessary work to be done, including the boarding up of openings, to render such structure temporarily safe whether or not the legal procedure herein described has been instituted; and shall cause such other action to be taken as the code official deems necessary to meet such emergency.

109.3 Closing streets. When necessary for public safety, the code official shall temporarily close structures and close, or order the authority having jurisdiction to close, sidewalks, streets, public ways and places adjacent to unsafe structures, and prohibit the same from being utilized.

109.4 Emergency repairs. For the purposes of this section, the code official shall employ the necessary labor and materials to perform the required work as expeditiously as possible.

109.5 Costs of emergency repairs. Costs incurred in the performance of emergency work shall be paid by the jurisdiction. The legal counsel of the jurisdiction shall institute appropriate action against the owner of the premises where the unsafe structure is or was located for the recovery of such costs.

109.6 Hearing. Any person ordered to take emergency measures shall comply with such order forthwith. Any affected person shall thereafter, upon petition directed to the appeals board, be afforded a hearing as described in this code.

SECTION 110
DEMOLITION

110.1 General. The code official shall order the owner of any premises upon which is located any structure, which in the code official's judgment is so old, dilapidated or has become so out of repair as to be dangerous, unsafe, insanitary or otherwise unfit for human habitation or occupancy, and such that it is unreasonable to repair the structure, to demolish and remove such structure; or if such structure is capable of being made safe by repairs, to repair and make safe and sanitary or to demolish and remove at the owner's option; or where there has been a cessation of normal construction of any structure for a period of more than two years, to demolish and remove such structure.

110.2 Notices and orders. All notices and orders shall comply with Section 107.

110.3 Failure to comply. If the owner of a premises fails to comply with a demolition order within the time prescribed, the code official shall cause the structure to be demolished and removed, either through an available public agency or by contract or arrangement with private persons, and the cost of such demolition and removal shall be charged against the real estate upon which the structure is located and shall be a lien upon such real estate.

110.4 Salvage materials. When any structure has been ordered demolished and removed, the governing body or other designated officer under said contract or arrangement aforesaid shall have the right to sell the salvage and valuable materials at the highest price obtainable. The net proceeds of such sale, after deducting the expenses of such demolition and removal, shall be promptly remitted with a report of such sale or transaction, including the items of expense and the amounts deducted, for the person who is entitled thereto, subject to any order of a court. If such a surplus does not remain to be turned over, the report shall so state.

SECTION 111
MEANS OF APPEAL

111.1 Application for appeal. Any person directly affected by a decision of the code official or a notice or order issued under this code shall have the right to appeal to the board of appeals, provided that a written application for appeal is filed within 20 days after the day the decision, notice or order was served. An application for appeal shall be based on a claim that the true intent of this code or the rules legally adopted thereunder have been incorrectly interpreted, the provisions of this code do not fully apply, or the requirements of this code are adequately satisfied by other means.

111.2 Membership of board. The board of appeals shall consist of a minimum of three members who are qualified by experience and training to pass on matters pertaining to property maintenance and who are not employees of the jurisdiction. The code official shall be an ex-officio member but shall have no vote on any matter before the board. The board shall be appointed by the chief appointing authority, and shall serve staggered and overlapping terms.

111.2.1 Alternate members. The chief appointing authority shall appoint two or more alternate members who shall be called by the board chairman to hear appeals during the absence or disqualification of a member. Alternate members shall possess the qualifications required for board membership.

111.2.2 Chairman. The board shall annually select one of its members to serve as chairman.

111.2.3 Disqualification of member. A member shall not hear an appeal in which that member has a personal, professional or financial interest.

111.2.4 Secretary. The chief administrative officer shall designate a qualified person to serve as secretary to the board. The secretary shall file a detailed record of all proceedings in the office of the chief administrative officer.

111.2.5 Compensation of members. Compensation of members shall be determined by law.

111.3 Notice of meeting. The board shall meet upon notice from the chairman, within 20 days of the filing of an appeal, or at stated periodic meetings.

111.4 Open hearing. All hearings before the board shall be open to the public. The appellant, the appellant's representative, the code official and any person whose interests are affected shall be given an opportunity to be heard. A quorum shall consist of not less than two-thirds of the board membership.

111.4.1 Procedure. The board shall adopt and make available to the public through the secretary procedures under which a hearing will be conducted. The procedures shall not require compliance with strict rules of evidence, but shall mandate that only relevant information be received.

111.5 Postponed hearing. When the full board is not present to hear an appeal, either the appellant or the appellant's representative shall have the right to request a postponement of the hearing.

111.6 Board decision. The board shall modify or reverse the decision of the code official only by a concurring vote of a majority of the total number of appointed board members.

111.6.1 Records and copies. The decision of the board shall be recorded. Copies shall be furnished to the appellant and to the code official.

111.6.2 Administration. The code official shall take immediate action in accordance with the decision of the board.

111.7 Court review. Any person, whether or not a previous party of the appeal, shall have the right to apply to the appropriate court for a writ of certiorari to correct errors of law. Application for review shall be made in the manner and time required by law following the filing of the decision in the office of the chief administrative officer.

111.8 Stays of enforcement. Appeals of notice and orders (other than Imminent Danger notices) shall stay the enforcement of the notice and order until the appeal is heard by the appeals board.

CHAPTER 2

DEFINITIONS

SECTION 201
GENERAL

201.1 Scope. Unless otherwise expressly stated, the following terms shall, for the purposes of this code, have the meanings shown in this chapter.

201.2 Interchangeability. Words stated in the present tense include the future; words stated in the masculine gender include the feminine and neuter; the singular number includes the plural and the plural, the singular.

201.3 Terms defined in other codes. Where terms are not defined in this code and are defined in the *International Building Code, International Fire Code, International Zoning Code, International Plumbing Code, International Mechanical Code* or the ICC *Electrical Code*, such terms shall have the meanings ascribed to them as stated in those codes.

201.4 Terms not defined. Where terms are not defined through the methods authorized by this section, such terms shall have ordinarily accepted meanings such as the context implies.

201.5 Parts. Whenever the words "dwelling unit," "dwelling," "premises," "building," "rooming house," "rooming unit" "housekeeping unit" or "story" are stated in this code, they shall be construed as though they were followed by the words "or any part thereof."

SECTION 202
GENERAL DEFINITIONS

APPROVED. Approved by the code official.

BASEMENT. That portion of a building which is partly or completely below grade.

BATHROOM. A room containing plumbing fixtures including a bathtub or shower.

BEDROOM. Any room or space used or intended to be used for sleeping purposes in either a dwelling or sleeping unit.

CODE OFFICIAL. The official who is charged with the administration and enforcement of this code, or any duly authorized representative.

CONDEMN. To adjudge unfit for occupancy.

[B] DWELLING UNIT. A single unit providing complete, independent living facilities for one or more persons, including permanent provisions for living, sleeping, eating, cooking and sanitation.

EASEMENT. That portion of land or property reserved for present or future use by a person or agency other than the legal fee owner(s) of the property. The easement shall be permitted to be for use under, on or above a said lot or lots.

EXTERIOR PROPERTY. The open space on the premises and on adjoining property under the control of owners or operators of such premises.

EXTERMINATION. The control and elimination of insects, rats or other pests by eliminating their harborage places; by removing or making inaccessible materials that serve as their food; by poison spraying, fumigating, trapping or by any other approved pest elimination methods.

GARBAGE. The animal or vegetable waste resulting from the handling, preparation, cooking and consumption of food.

GUARD. A building component or a system of building components located at or near the open sides of elevated walking surfaces that minimizes the possibility of a fall from the walking surface to a lower level.

HABITABLE SPACE. Space in a structure for living, sleeping, eating or cooking. Bathrooms, toilet rooms, closets, halls, storage or utility spaces, and similar areas are not considered habitable spaces.

HOUSEKEEPING UNIT. A room or group of rooms forming a single habitable space equipped and intended to be used for living, sleeping, cooking and eating which does not contain, within such a unit, a toilet, lavatory and bathtub or shower.

IMMINENT DANGER. A condition which could cause serious or life-threatening injury or death at any time.

INFESTATION. The presence, within or contiguous to, a structure or premises of insects, rats, vermin or other pests.

INOPERABLE MOTOR VEHICLE. A vehicle which cannot be driven upon the public streets for reason including but not limited to being unlicensed, wrecked, abandoned, in a state of disrepair, or incapable of being moved under its own power.

LABELED. Devices, equipment, appliances, or materials to which has been affixed a label, seal, symbol or other identifying mark of a nationally recognized testing laboratory, inspection agency or other organization concerned with product evaluation that maintains periodic inspection of the production of the above-labeled items and by whose label the manufacturer attests to compliance with applicable nationally recognized standards.

LET FOR OCCUPANCY OR LET. To permit, provide or offer possession or occupancy of a dwelling, dwelling unit, rooming unit, building, premise or structure by a person who is or is not the legal owner of record thereof, pursuant to a written or unwritten lease, agreement or license, or pursuant to a recorded or unrecorded agreement of contract for the sale of land.

OCCUPANCY. The purpose for which a building or portion thereof is utilized or occupied.

OCCUPANT. Any individual living or sleeping in a building, or having possession of a space within a building.

OPENABLE AREA. That part of a window, skylight or door which is available for unobstructed ventilation and which opens directly to the outdoors.

OPERATOR. Any person who has charge, care or control of a structure or premises which is let or offered for occupancy.

OWNER. Any person, agent, operator, firm or corporation having a legal or equitable interest in the property; or recorded in the official records of the state, county or municipality as holding title to the property; or otherwise having control of the property, including the guardian of the estate of any such person, and the executor or administrator of the estate of such person if ordered to take possession of real property by a court.

PERSON. An individual, corporation, partnership or any other group acting as a unit.

PREMISES. A lot, plot or parcel of land, easement or public way, including any structures thereon.

PUBLIC WAY. Any street, alley or similar parcel of land essentially unobstructed from the ground to the sky, which is deeded, dedicated or otherwise permanently appropriated to the public for public use.

ROOMING HOUSE. A building arranged or occupied for lodging, with or without meals, for compensation and not occupied as a one- or two-family dwelling.

ROOMING UNIT. Any room or group of rooms forming a single habitable unit occupied or intended to be occupied for sleeping or living, but not for cooking purposes.

RUBBISH. Combustible and noncombustible waste materials, except garbage; the term shall include the residue from the burning of wood, coal, coke and other combustible materials, paper, rags, cartons, boxes, wood, excelsior, rubber, leather, tree branches, yard trimmings, tin cans, metals, mineral matter, glass, crockery and dust and other similar materials.

[B] SLEEPING UNIT. A room or space in which people sleep, which can also include permanent provisions for living, eating and either sanitation or kitchen facilities, but not both. Such rooms and spaces that are also part of a dwelling unit are not sleeping units.

STRICT LIABILITY OFFENSE. An offense in which the prosecution in a legal proceeding is not required to prove criminal intent as a part of its case. It is enough to prove that the defendant either did an act which was prohibited, or failed to do an act which the defendant was legally required to do.

STRUCTURE. That which is built or constructed or a portion thereof.

TENANT. A person, corporation, partnership or group, whether or not the legal owner of record, occupying a building or portion thereof as a unit.

TOILET ROOM. A room containing a water closet or urinal but not a bathtub or shower.

VENTILATION. The natural or mechanical process of supplying conditioned or unconditioned air to, or removing such air from, any space.

WORKMANLIKE. Executed in a skilled manner; e.g., generally plumb, level, square, in line, undamaged and without marring adjacent work.

YARD. An open space on the same lot with a structure.

CHAPTER 3

GENERAL REQUIREMENTS

SECTION 301
GENERAL

301.1 Scope. The provisions of this chapter shall govern the minimum conditions and the responsibilities of persons for maintenance of structures, equipment and exterior property.

301.2 Responsibility. The owner of the premises shall maintain the structures and exterior property in compliance with these requirements, except as otherwise provided for in this code. A person shall not occupy as owner-occupant or permit another person to occupy premises which are not in a sanitary and safe condition and which do not comply with the requirements of this chapter. Occupants of a dwelling unit, rooming unit or housekeeping unit are responsible for keeping in a clean, sanitary and safe condition that part of the dwelling unit, rooming unit, housekeeping unit or premises which they occupy and control.

301.3 Vacant structures and land. All vacant structures and premises thereof or vacant land shall be maintained in a clean, safe, secure and sanitary condition as provided herein so as not to cause a blighting problem or adversely affect the public health or safety.

SECTION 302
EXTERIOR PROPERTY AREAS

302.1 Sanitation. All exterior property and premises shall be maintained in a clean, safe and sanitary condition. The occupant shall keep that part of the exterior property which such occupant occupies or controls in a clean and sanitary condition.

302.2 Grading and drainage. All premises shall be graded and maintained to prevent the erosion of soil and to prevent the accumulation of stagnant water thereon, or within any structure located thereon.

Exception: Approved retention areas and reservoirs.

302.3 Sidewalks and driveways. All sidewalks, walkways, stairs, driveways, parking spaces and similar areas shall be kept in a proper state of repair, and maintained free from hazardous conditions.

302.4 Weeds. All premises and exterior property shall be maintained free from weeds or plant growth in excess of (jurisdiction to insert height in inches). All noxious weeds shall be prohibited. Weeds shall be defined as all grasses, annual plants and vegetation, other than trees or shrubs provided; however, this term shall not include cultivated flowers and gardens.

Upon failure of the owner or agent having charge of a property to cut and destroy weeds after service of a notice of violation, they shall be subject to prosecution in accordance with Section 106.3 and as prescribed by the authority having jurisdiction. Upon failure to comply with the notice of violation, any duly authorized employee of the jurisdiction or contractor hired by the jurisdiction shall be authorized to enter upon the property in violation and cut and destroy the weeds growing thereon, and the costs of such removal shall be paid by the owner or agent responsible for the property.

302.5 Rodent harborage. All structures and exterior property shall be kept free from rodent harborage and infestation. Where rodents are found, they shall be promptly exterminated by approved processes which will not be injurious to human health. After extermination, proper precautions shall be taken to eliminate rodent harborage and prevent reinfestation.

302.6 Exhaust vents. Pipes, ducts, conductors, fans or blowers shall not discharge gases, steam, vapor, hot air, grease, smoke, odors or other gaseous or particulate wastes directly upon abutting or adjacent public or private property or that of another tenant.

302.7 Accessory structures. All accessory structures, including detached garages, fences and walls, shall be maintained structurally sound and in good repair.

302.8 Motor vehicles. Except as provided for in other regulations, no inoperative or unlicensed motor vehicle shall be parked, kept or stored on any premises, and no vehicle shall at any time be in a state of major disassembly, disrepair, or in the process of being stripped or dismantled. Painting of vehicles is prohibited unless conducted inside an approved spray booth.

Exception: A vehicle of any type is permitted to undergo major overhaul, including body work, provided that such work is performed inside a structure or similarly enclosed area designed and approved for such purposes.

302.9 Defacement of property. No person shall willfully or wantonly damage, mutilate or deface any exterior surface of any structure or building on any private or public property by placing thereon any marking, carving or graffiti.

It shall be the responsibility of the owner to restore said surface to an approved state of maintenance and repair.

SECTION 303
SWIMMING POOLS, SPAS AND HOT TUBS

303.1 Swimming pools. Swimming pools shall be maintained in a clean and sanitary condition, and in good repair.

303.2 Enclosures. Private swimming pools, hot tubs and spas, containing water more than 24 inches (610 mm) in depth shall be completely surrounded by a fence or barrier at least 48 inches (1219 mm) in height above the finished ground level measured on the side of the barrier away from the pool. Gates and doors in such barriers shall be self-closing and self-latching. Where the self-latching device is less than 54 inches (1372 mm) above the bottom of the gate, the release mechanism shall be located on the pool side of the gate. Self-closing and self-latching gates shall be maintained such that the gate will positively close and latch when released from an open position of 6 inches (152 mm) from the gatepost. No existing pool enclosure shall be removed, replaced or changed in a manner that reduces its effectiveness as a safety barrier.

Exception: Spas or hot tubs with a safety cover that complies with ASTM F 1346 shall be exempt from the provisions of this section.

SECTION 304
EXTERIOR STRUCTURE

304.1 General. The exterior of a structure shall be maintained in good repair, structurally sound and sanitary so as not to pose a threat to the public health, safety or welfare.

304.2 Protective treatment. All exterior surfaces, including but not limited to, doors, door and window frames, cornices, porches, trim, balconies, decks and fences shall be maintained in good condition. Exterior wood surfaces, other than decay-resistant woods, shall be protected from the elements and decay by painting or other protective covering or treatment. Peeling, flaking and chipped paint shall be eliminated and surfaces repainted. All siding and masonry joints as well as those between the building envelope and the perimeter of windows, doors, and skylights shall be maintained weather resistant and water tight. All metal surfaces subject to rust or corrosion shall be coated to inhibit such rust and corrosion and all surfaces with rust or corrosion shall be stabilized and coated to inhibit future rust and corrosion. Oxidation stains shall be removed from exterior surfaces. Surfaces designed for stabilization by oxidation are exempt from this requirement.

[F] 304.3 Premises identification. Buildings shall have approved address numbers placed in a position to be plainly legible and visible from the street or road fronting the property. These numbers shall contrast with their background. Address numbers shall be Arabic numerals or alphabet letters. Numbers shall be a minimum of 4 inches (102 mm) high with a minimum stroke width of 0.5 inch (12.7 mm).

304.4 Structural members. All structural members shall be maintained free from deterioration, and shall be capable of safely supporting the imposed dead and live loads.

304.5 Foundation walls. All foundation walls shall be maintained plumb and free from open cracks and breaks and shall be kept in such condition so as to prevent the entry of rodents and other pests.

304.6 Exterior walls. All exterior walls shall be free from holes, breaks, and loose or rotting materials; and maintained weatherproof and properly surface coated where required to prevent deterioration.

304.7 Roofs and drainage. The roof and flashing shall be sound, tight and not have defects that admit rain. Roof drainage shall be adequate to prevent dampness or deterioration in the walls or interior portion of the structure. Roof drains, gutters and downspouts shall be maintained in good repair and free from obstructions. Roof water shall not be discharged in a manner that creates a public nuisance.

304.8 Decorative features. All cornices, belt courses, corbels, terra cotta trim, wall facings and similar decorative features shall be maintained in good repair with proper anchorage and in a safe condition.

304.9 Overhang extensions. All overhang extensions including, but not limited to canopies, marquees, signs, metal awnings, fire escapes, standpipes and exhaust ducts shall be maintained in good repair and be properly anchored so as to be kept in a sound condition. When required, all exposed surfaces of metal or wood shall be protected from the elements and against decay or rust by periodic application of weather-coating materials, such as paint or similar surface treatment.

304.10 Stairways, decks, porches and balconies. Every exterior stairway, deck, porch and balcony, and all appurtenances attached thereto, shall be maintained structurally sound, in good repair, with proper anchorage and capable of supporting the imposed loads.

304.11 Chimneys and towers. All chimneys, cooling towers, smoke stacks, and similar appurtenances shall be maintained structurally safe and sound, and in good repair. All exposed surfaces of metal or wood shall be protected from the elements and against decay or rust by periodic application of weather-coating materials, such as paint or similar surface treatment.

304.12 Handrails and guards. Every handrail and guard shall be firmly fastened and capable of supporting normally imposed loads and shall be maintained in good condition.

304.13 Window, skylight and door frames. Every window, skylight, door and frame shall be kept in sound condition, good repair and weather tight.

304.13.1 Glazing. All glazing materials shall be maintained free from cracks and holes.

304.13.2 Openable windows. Every window, other than a fixed window, shall be easily openable and capable of being held in position by window hardware.

304.14 Insect screens. During the period from [DATE] to [DATE], every door, window and other outside opening required for ventilation of habitable rooms, food preparation areas, food service areas or any areas where products to be included or utilized in food for human consumption are processed, manufactured, packaged or stored shall be supplied with approved tightly fitting screens of not less than 16 mesh per inch (16 mesh per 25 mm), and every screen door used for insect control shall have a self-closing device in good working condition.

Exception: Screens shall not be required where other approved means, such as air curtains or insect repellent fans, are employed.

304.15 Doors. All exterior doors, door assemblies and hardware shall be maintained in good condition. Locks at all entrances to dwelling units and sleeping units shall tightly secure the door. Locks on means of egress doors shall be in accordance with Section 702.3.

304.16 Basement hatchways. Every basement hatchway shall be maintained to prevent the entrance of rodents, rain and surface drainage water.

304.17 Guards for basement windows. Every basement window that is openable shall be supplied with rodent shields, storm windows or other approved protection against the entry of rodents.

304.18 Building security. Doors, windows or hatchways for dwelling units, room units or housekeeping units shall be provided with devices designed to provide security for the occupants and property within.

304.18.1 Doors. Doors providing access to a dwelling unit, rooming unit or housekeeping unit that is rented, leased or let shall be equipped with a deadbolt lock designed to be readily openable from the side from which egress is to be made without the need for keys, special knowledge or effort and shall have a lock throw of not less than 1 inch (25 mm). Such deadbolt locks shall be installed according to the manufacturer's specifications and maintained in good working order. For the purpose of this section, a sliding bolt shall not be considered an acceptable deadbolt lock.

304.18.2 Windows. Operable windows located in whole or in part within 6 feet (1828 mm) above ground level or a walking surface below that provide access to a dwelling unit, rooming unit or housekeeping unit that is rented, leased or let shall be equipped with a window sash locking device.

304.18.3 Basement hatchways. Basement hatchways that provide access to a dwelling unit, rooming unit or housekeeping unit that is rented, leased or let shall be equipped with devices that secure the units from unauthorized entry.

SECTION 305
INTERIOR STRUCTURE

305.1 General. The interior of a structure and equipment therein shall be maintained in good repair, structurally sound and in a sanitary condition. Occupants shall keep that part of the structure which they occupy or control in a clean and sanitary condition. Every owner of a structure containing a rooming house, housekeeping units, a hotel, a dormitory, two or more dwelling units or two or more nonresidential occupancies, shall maintain, in a clean and sanitary condition, the shared or public areas of the structure and exterior property.

305.2 Structural members. All structural members shall be maintained structurally sound, and be capable of supporting the imposed loads.

305.3 Interior surfaces. All interior surfaces, including windows and doors, shall be maintained in good, clean and sanitary condition. Peeling, chipping, flaking or abraded paint shall be repaired, removed or covered. Cracked or loose plaster, decayed wood and other defective surface conditions shall be corrected.

305.4 Stairs and walking surfaces. Every stair, ramp, landing, balcony, porch, deck or other walking surface shall be maintained in sound condition and good repair.

305.5 Handrails and guards. Every handrail and guard shall be firmly fastened and capable of supporting normally imposed loads and shall be maintained in good condition.

305.6 Interior doors. Every interior door shall fit reasonably well within its frame and shall be capable of being opened and closed by being properly and securely attached to jambs, headers or tracks as intended by the manufacturer of the attachment hardware.

SECTION 306
HANDRAILS AND GUARDRAILS

306.1 General. Every exterior and interior flight of stairs having more than four risers shall have a handrail on one side of the stair and every open portion of a stair, landing, balcony, porch, deck, ramp or other walking surface which is more than 30 inches (762 mm) above the floor or grade below shall have guards. Handrails shall not be less than 30 inches (762 mm) high or more than 42 inches (1067 mm) high measured vertically above the nosing of the tread or above the finished floor of the landing or walking surfaces. Guards shall not be less than 30 inches (762 mm) high above the floor of the landing, balcony, porch, deck, or ramp or other walking surface.

Exception: Guards shall not be required where exempted by the adopted building code.

SECTION 307
RUBBISH AND GARBAGE

307.1 Accumulation of rubbish or garbage. All exterior property and premises, and the interior of every structure, shall be free from any accumulation of rubbish or garbage.

307.2 Disposal of rubbish. Every occupant of a structure shall dispose of all rubbish in a clean and sanitary manner by placing such rubbish in approved containers.

307.2.1 Rubbish storage facilities. The owner of every occupied premises shall supply approved covered containers for rubbish, and the owner of the premises shall be responsible for the removal of rubbish.

307.2.2 Refrigerators. Refrigerators and similar equipment not in operation shall not be discarded, abandoned or stored on premises without first removing the doors.

307.3 Disposal of garbage. Every occupant of a structure shall dispose of garbage in a clean and sanitary manner by placing such garbage in an approved garbage disposal facility or approved garbage containers.

307.3.1 Garbage facilities. The owner of every dwelling shall supply one of the following: an approved mechanical food waste grinder in each dwelling unit; an approved incinerator unit in the structure available to the occupants in each dwelling unit; or an approved leakproof, covered, outside garbage container.

307.3.2 Containers. The operator of every establishment producing garbage shall provide, and at all times cause to be utilized, approved leakproof containers provided with close-fitting covers for the storage of such materials until removed from the premises for disposal.

SECTION 308
EXTERMINATION

308.1 Infestation. All structures shall be kept free from insect and rodent infestation. All structures in which insects or rodents are found shall be promptly exterminated by approved processes that will not be injurious to human health. After extermination, proper precautions shall be taken to prevent reinfestation.

308.2 Owner. The owner of any structure shall be responsible for extermination within the structure prior to renting or leasing the structure.

308.3 Single occupant. The occupant of a one-family dwelling or of a single-tenant nonresidential structure shall be responsible for extermination on the premises.

308.4 Multiple occupancy. The owner of a structure containing two or more dwelling units, a multiple occupancy, a rooming house or a nonresidential structure shall be responsible for extermination in the public or shared areas of the structure and exterior property. If infestation is caused by failure of an occupant to prevent such infestation in the area occupied, the occupant shall be responsible for extermination.

308.5 Occupant. The occupant of any structure shall be responsible for the continued rodent and pest-free condition of the structure.

> **Exception:** Where the infestations are caused by defects in the structure, the owner shall be responsible for extermination.

CHAPTER 4

LIGHT, VENTILATION AND OCCUPANCY LIMITATIONS

SECTION 401
GENERAL

401.1 Scope. The provisions of this chapter shall govern the minimum conditions and standards for light, ventilation and space for occupying a structure.

401.2 Responsibility. The owner of the structure shall provide and maintain light, ventilation and space conditions in compliance with these requirements. A person shall not occupy as owner-occupant, or permit another person to occupy, any premises that do not comply with the requirements of this chapter.

401.3 Alternative devices. In lieu of the means for natural light and ventilation herein prescribed, artificial light or mechanical ventilation complying with the *International Building Code* shall be permitted.

SECTION 402
LIGHT

402.1 Habitable spaces. Every habitable space shall have at least one window of approved size facing directly to the outdoors or to a court. The minimum total glazed area for every habitable space shall be 8 percent of the floor area of such room. Wherever walls or other portions of a structure face a window of any room and such obstructions are located less than 3 feet (914 mm) from the window and extend to a level above that of the ceiling of the room, such window shall not be deemed to face directly to the outdoors nor to a court and shall not be included as contributing to the required minimum total window area for the room.

> **Exception:** Where natural light for rooms or spaces without exterior glazing areas is provided through an adjoining room, the unobstructed opening to the adjoining room shall be at least 8 percent of the floor area of the interior room or space, but not less than 25 square feet (2.33 m²). The exterior glazing area shall be based on the total floor area being served.

402.2 Common halls and stairways. Every common hall and stairway in residential occupancies, other than in one- and two-family dwellings, shall be lighted at all times with at least a 60-watt standard incandescent light bulb for each 200 square feet (19 m²) of floor area or equivalent illumination, provided that the spacing between lights shall not be greater than 30 feet (9144 mm). In other than residential occupancies, means of egress, including exterior means of egress, stairways shall be illuminated at all times the building space served by the means of egress is occupied with a minimum of 1 footcandle (11 lux) at floors, landings and treads.

402.3 Other spaces. All other spaces shall be provided with natural or artificial light sufficient to permit the maintenance of sanitary conditions, and the safe occupancy of the space and utilization of the appliances, equipment and fixtures.

SECTION 403
VENTILATION

403.1 Habitable spaces. Every habitable space shall have at least one openable window. The total openable area of the window in every room shall be equal to at least 45 percent of the minimum glazed area required in Section 402.1.

> **Exception:** Where rooms and spaces without openings to the outdoors are ventilated through an adjoining room, the unobstructed opening to the adjoining room shall be at least 8 percent of the floor area of the interior room or space, but not less than 25 square feet (2.33 m²). The ventilation openings to the outdoors shall be based on a total floor area being ventilated.

403.2 Bathrooms and toilet rooms. Every bathroom and toilet room shall comply with the ventilation requirements for habitable spaces as required by Section 403.1, except that a window shall not be required in such spaces equipped with a mechanical ventilation system. Air exhausted by a mechanical ventilation system from a bathroom or toilet room shall discharge to the outdoors and shall not be recirculated.

403.3 Cooking facilities. Unless approved through the certificate of occupancy, cooking shall not be permitted in any rooming unit or dormitory unit, and a cooking facility or appliance shall not be permitted to be present in the rooming unit or dormitory unit.

> **Exceptions:**
>
> 1. Where specifically approved in writing by the code official.
>
> 2. Devices such as coffee pots and microwave ovens shall not be considered cooking appliances.

403.4 Process ventilation. Where injurious, toxic, irritating or noxious fumes, gases, dusts or mists are generated, a local exhaust ventilation system shall be provided to remove the contaminating agent at the source. Air shall be exhausted to the exterior and not be recirculated to any space.

403.5 Clothes dryer exhaust. Clothes dryer exhaust systems shall be independent of all other systems and shall be exhausted in accordance with the manufacturer's instructions.

SECTION 404
OCCUPANCY LIMITATIONS

404.1 Privacy. Dwelling units, hotel units, housekeeping units, rooming units and dormitory units shall be arranged to provide privacy and be separate from other adjoining spaces.

404.2 Minimum room widths. A habitable room, other than a kitchen, shall not be less than 7 feet (2134 mm) in any plan dimension. Kitchens shall have a clear passageway of not less than 3 feet (914 mm) between counterfronts and appliances or counterfronts and walls.

404.3 Minimum ceiling heights. Habitable spaces, hallways, corridors, laundry areas, bathrooms, toilet rooms and habitable basement areas shall have a clear ceiling height of not less than 7 feet (2134 mm).

Exceptions:

1. In one- and two-family dwellings, beams or girders spaced not less than 4 feet (1219 mm) on center and projecting not more than 6 inches (152 mm) below the required ceiling height.

2. Basement rooms in one- and two-family dwellings occupied exclusively for laundry, study or recreation purposes, having a ceiling height of not less than 6 feet 8 inches (2033 mm) with not less than 6 feet 4 inches (1932 mm) of clear height under beams, girders, ducts and similar obstructions.

3. Rooms occupied exclusively for sleeping, study or similar purposes and having a sloped ceiling over all or part of the room, with a clear ceiling height of at least 7 feet (2134 mm) over not less than one-third of the required minimum floor area. In calculating the floor area of such rooms, only those portions of the floor area with a clear ceiling height of 5 feet (1524 mm) or more shall be included.

404.4 Bedroom and living room requirements. Every bedroom and living room shall comply with the requirements of Sections 404.4.1 through 404.4.5.

404.4.1 Room area. Every living room shall contain at least 120 square feet (11.2 m²) and every bedroom shall contain at least 70 square feet (6.5 m²).

404.4.2 Access from bedrooms. Bedrooms shall not constitute the only means of access to other bedrooms or habitable spaces and shall not serve as the only means of egress from other habitable spaces.

Exception: Units that contain fewer than two bedrooms.

404.4.3 Water closet accessibility. Every bedroom shall have access to at least one water closet and one lavatory without passing through another bedroom. Every bedroom in a dwelling unit shall have access to at least one water closet and lavatory located in the same story as the bedroom or an adjacent story.

404.4.4 Prohibited occupancy. Kitchens and nonhabitable spaces shall not be used for sleeping purposes.

404.4.5 Other requirements. Bedrooms shall comply with the applicable provisions of this code including, but not limited to, the light, ventilation, room area, ceiling height and room width requirements of this chapter; the plumbing facilities and water-heating facilities requirements of Chapter 5; the heating facilities and electrical receptacle requirements of Chapter 6; and the smoke detector and emergency escape requirements of Chapter 7.

404.5 Overcrowding. The number of persons occupying a dwelling unit shall not create conditions that, in the opinion of the code official, endanger the life, health, safety or welfare of the occupants.

404.6 Efficiency unit. Nothing in this section shall prohibit an efficiency living unit from meeting the following requirements:

1. A unit occupied by not more than two occupants shall have a clear floor area of not less than 220 square feet (20.4 m²). A unit occupied by three occupants shall have a clear floor area of not less than 320 square feet (29.7 m²). These required areas shall be exclusive of the areas required by Items 2 and 3.

2. The unit shall be provided with a kitchen sink, cooking appliance and refrigeration facilities, each having a clear working space of not less than 30 inches (762 mm) in front. Light and ventilation conforming to this code shall be provided.

3. The unit shall be provided with a separate bathroom containing a water closet, lavatory and bathtub or shower.

4. The maximum number of occupants shall be three.

404.7 Food preparation. All spaces to be occupied for food preparation purposes shall contain suitable space and equipment to store, prepare and serve foods in a sanitary manner. There shall be adequate facilities and services for the sanitary disposal of food wastes and refuse, including facilities for temporary storage.

CHAPTER 5

PLUMBING FACILITIES AND FIXTURE REQUIREMENTS

SECTION 501
GENERAL

501.1 Scope. The provisions of this chapter shall govern the minimum plumbing systems, facilities and plumbing fixtures to be provided.

501.2 Responsibility. The owner of the structure shall provide and maintain such plumbing facilities and plumbing fixtures in compliance with these requirements. A person shall not occupy as owner-occupant or permit another person to occupy any structure or premises which does not comply with the requirements of this chapter.

[P] SECTION 502
REQUIRED FACILITIES

502.1 Dwelling units. Every dwelling unit shall contain its own bathtub or shower, lavatory, water closet and kitchen sink which shall be maintained in a sanitary, safe working condition. The lavatory shall be placed in the same room as the water closet or located in close proximity to the door leading directly into the room in which such water closet is located. A kitchen sink shall not be used as a substitute for the required lavatory.

502.2 Rooming houses. At least one water closet, lavatory and bathtub or shower shall be supplied for each four rooming units.

502.3 Hotels. Where private water closets, lavatories and baths are not provided, one water closet, one lavatory and one bathtub or shower having access from a public hallway shall be provided for each ten occupants.

502.4 Employees' facilities. A minimum of one water closet, one lavatory and one drinking facility shall be available to employees.

>   **502.4.1 Drinking facilities.** Drinking facilities shall be a drinking fountain, water cooler, bottled water cooler or disposable cups next to a sink or water dispenser. Drinking facilities shall not be located in toilet rooms or bathrooms.

[P] SECTION 503
TOILET ROOMS

503.1 Privacy. Toilet rooms and bathrooms shall provide privacy and shall not constitute the only passageway to a hall or other space, or to the exterior. A door and interior locking device shall be provided for all common or shared bathrooms and toilet rooms in a multiple dwelling.

503.2 Location. Toilet rooms and bathrooms serving hotel units, rooming units or dormitory units or housekeeping units, shall have access by traversing not more than one flight of stairs and shall have access from a common hall or passageway.

503.3 Location of employee toilet facilities. Toilet facilities shall have access from within the employees' working area. The required toilet facilities shall be located not more than one story above or below the employees' working area and the path of travel to such facilities shall not exceed a distance of 500 feet (152 m). Employee facilities shall either be separate facilities or combined employee and public facilities.

>   **Exception:** Facilities that are required for employees in storage structures or kiosks, which are located in adjacent structures under the same ownership, lease or control, shall not exceed a travel distance of 500 feet (152 m) from the employees' regular working area to the facilities.

503.4 Floor surface. In other than dwelling units, every toilet room floor shall be maintained to be a smooth, hard, nonabsorbent surface to permit such floor to be easily kept in a clean and sanitary condition.

[P] SECTION 504
PLUMBING SYSTEMS AND FIXTURES

504.1 General. All plumbing fixtures shall be properly installed and maintained in working order, and shall be kept free from obstructions, leaks and defects and be capable of performing the function for which such plumbing fixtures are designed. All plumbing fixtures shall be maintained in a safe, sanitary and functional condition.

504.2 Fixture clearances. Plumbing fixtures shall have adequate clearances for usage and cleaning.

504.3 Plumbing system hazards. Where it is found that a plumbing system in a structure constitutes a hazard to the occupants or the structure by reason of inadequate service, inadequate venting, cross connection, backsiphonage, improper installation, deterioration or damage or for similar reasons, the code official shall require the defects to be corrected to eliminate the hazard.

SECTION 505
WATER SYSTEM

505.1 General. Every sink, lavatory, bathtub or shower, drinking fountain, water closet or other plumbing fixture shall be properly connected to either a public water system or to an approved private water system. All kitchen sinks, lavatories, laundry facilities, bathtubs and showers shall be supplied with hot or tempered and cold running water in accordance with the *International Plumbing Code.*

[P] 505.2 Contamination. The water supply shall be maintained free from contamination, and all water inlets for plumbing fixtures shall be located above the flood-level rim of the fixture. Shampoo basin faucets, janitor sink faucets and other hose bibs or faucets to which hoses are attached and left in place, shall be protected by an approved atmospheric-type vacuum breaker or an approved permanently attached hose connection vacuum breaker.

505.3 Supply. The water supply system shall be installed and maintained to provide a supply of water to plumbing fixtures, devices and appurtenances in sufficient volume and at pressures adequate to enable the fixtures to function properly, safely, and free from defects and leaks.

505.4 Water heating facilities. Water heating facilities shall be properly installed, maintained and capable of providing an adequate amount of water to be drawn at every required sink, lavatory, bathtub, shower and laundry facility at a temperature of not less than 110°F (43°C). A gas-burning water heater shall not be located in any bathroom, toilet room, bedroom or other occupied room normally kept closed, unless adequate combustion air is provided. An approved combination temperature and pressure-relief valve and relief valve discharge pipe shall be properly installed and maintained on water heaters.

[P] SECTION 506
SANITARY DRAINAGE SYSTEM

506.1 General. All plumbing fixtures shall be properly connected to either a public sewer system or to an approved private sewage disposal system.

506.2 Maintenance. Every plumbing stack, vent, waste and sewer line shall function properly and be kept free from obstructions, leaks and defects.

[P] SECTION 507
STORM DRAINAGE

507.1 General. Drainage of roofs and paved areas, yards and courts, and other open areas on the premises shall not be discharged in a manner that creates a public nuisance.

CHAPTER 6

MECHANICAL AND ELECTRICAL REQUIREMENTS

SECTION 601
GENERAL

601.1 Scope. The provisions of this chapter shall govern the minimum mechanical and electrical facilities and equipment to be provided.

601.2 Responsibility. The owner of the structure shall provide and maintain mechanical and electrical facilities and equipment in compliance with these requirements. A person shall not occupy as owner-occupant or permit another person to occupy any premises which does not comply with the requirements of this chapter.

SECTION 602
HEATING FACILITIES

602.1 Facilities required. Heating facilities shall be provided in structures as required by this section.

602.2 Residential occupancies. Dwellings shall be provided with heating facilities capable of maintaining a room temperature of 68°F (20°C) in all habitable rooms, bathrooms and toilet rooms based on the winter outdoor design temperature for the locality indicated in Appendix D of the *International Plumbing Code*. Cooking appliances shall not be used to provide space heating to meet the requirements of this section.

> **Exception:** In areas where the average monthly temperature is above 30°F (-1°C), a minimum temperature of 65°F (18°C) shall be maintained.

602.3 Heat supply. Every owner and operator of any building who rents, leases or lets one or more dwelling units or sleeping units on terms, either expressed or implied, to furnish heat to the occupants thereof shall supply heat during the period from [DATE] to [DATE] to maintain a temperature of not less than 68°F (20°C) in all habitable rooms, bathrooms, and toilet rooms.

> **Exceptions:**
> 1. When the outdoor temperature is below the winter outdoor design temperature for the locality, maintenance of the minimum room temperature shall not be required provided that the heating system is operating at its full design capacity. The winter outdoor design temperature for the locality shall be as indicated in Appendix D of the *International Plumbing Code*.
> 2. In areas where the average monthly temperature is above 30°F (-1°C) a minimum temperature of 65°F (18°C) shall be maintained.

602.4 Occupiable work spaces. Indoor occupiable work spaces shall be supplied with heat during the period from [DATE] to [DATE] to maintain a temperature of not less than 65°F (18°C) during the period the spaces are occupied.

> **Exceptions:**
> 1. Processing, storage and operation areas that require cooling or special temperature conditions.
> 2. Areas in which persons are primarily engaged in vigorous physical activities.

602.5 Room temperature measurement. The required room temperatures shall be measured 3 feet (914 mm) above the floor near the center of the room and 2 feet (610 mm) inward from the center of each exterior wall.

SECTION 603
MECHANICAL EQUIPMENT

603.1 Mechanical appliances. All mechanical appliances, fireplaces, solid fuel-burning appliances, cooking appliances and water heating appliances shall be properly installed and maintained in a safe working condition, and shall be capable of performing the intended function.

603.2 Removal of combustion products. All fuel-burning equipment and appliances shall be connected to an approved chimney or vent.

> **Exception:** Fuel-burning equipment and appliances which are labeled for unvented operation.

603.3 Clearances. All required clearances to combustible materials shall be maintained.

603.4 Safety controls. All safety controls for fuel-burning equipment shall be maintained in effective operation.

603.5 Combustion air. A supply of air for complete combustion of the fuel and for ventilation of the space containing the fuel-burning equipment shall be provided for the fuel-burning equipment.

603.6 Energy conservation devices. Devices intended to reduce fuel consumption by attachment to a fuel-burning appliance, to the fuel supply line thereto, or to the vent outlet or vent piping therefrom, shall not be installed unless labeled for such purpose and the installation is specifically approved.

SECTION 604
ELECTRICAL FACILITIES

604.1 Facilities required. Every occupied building shall be provided with an electrical system in compliance with the requirements of this section and Section 605.

604.2 Service. The size and usage of appliances and equipment shall serve as a basis for determining the need for additional facilities in accordance with the ICC *Electrical Code*. Dwelling units shall be served by a three-wire, 120/240 volt, single-phase electrical service having a rating of not less than 60 amperes.

604.3 Electrical system hazards. Where it is found that the electrical system in a structure constitutes a hazard to the occupants or the structure by reason of inadequate service, improper fusing, insufficient receptacle and lighting outlets, improper wiring or installation, deterioration or damage, or for similar reasons, the code official shall require the defects to be corrected to eliminate the hazard.

SECTION 605
ELECTRICAL EQUIPMENT

605.1 Installation. All electrical equipment, wiring and appliances shall be properly installed and maintained in a safe and approved manner.

605.2 Receptacles. Every habitable space in a dwelling shall contain at least two separate and remote receptacle outlets. Every laundry area shall contain at least one grounded-type receptacle or a receptacle with a ground fault circuit interrupter. Every bathroom shall contain at least one receptacle. Any new bathroom receptacle outlet shall have ground fault circuit interrupter protection.

605.3 Luminaires. Every public hall, interior stairway, toilet room, kitchen, bathroom, laundry room, boiler room and furnace room shall contain at least one electric luminaire.

SECTION 606
ELEVATORS, ESCALATORS AND DUMBWAITERS

606.1 General. Elevators, dumbwaiters and escalators shall be maintained in compliance with ASME A17.1. The most current certification of inspection shall be on display at all times within the elevator or attached to the escalator or dumbwaiter, or the certificate shall be available for public inspection in the office of the building operator. The inspection and tests shall be performed at not less than the periodical intervals listed in ASME A17.1, Appendix N, except where otherwise specified by the authority having jurisdiction.

606.2 Elevators. In buildings equipped with passenger elevators, at least one elevator shall be maintained in operation at all times when the building is occupied.

> **Exception:** Buildings equipped with only one elevator shall be permitted to have the elevator temporarily out of service for testing or servicing.

SECTION 607
DUCT SYSTEMS

607.1 General. Duct systems shall be maintained free of obstructions and shall be capable of performing the required function.

CHAPTER 7

FIRE SAFETY REQUIREMENTS

SECTION 701
GENERAL

701.1 Scope. The provisions of this chapter shall govern the minimum conditions and standards for fire safety relating to structures and exterior premises, including fire safety facilities and equipment to be provided.

701.2 Responsibility. The owner of the premises shall provide and maintain such fire safety facilities and equipment in compliance with these requirements. A person shall not occupy as owner-occupant or permit another person to occupy any premises that do not comply with the requirements of this chapter.

[F] SECTION 702
MEANS OF EGRESS

702.1 General. A safe, continuous and unobstructed path of travel shall be provided from any point in a building or structure to the public way. Means of egress shall comply with the *International Fire Code.*

702.2 Aisles. The required width of aisles in accordance with the *International Fire Code* shall be unobstructed.

702.3 Locked doors. All means of egress doors shall be readily openable from the side from which egress is to be made without the need for keys, special knowledge or effort, except where the door hardware conforms to that permitted by the *International Building Code.*

702.4 Emergency escape openings. Required emergency escape openings shall be maintained in accordance with the code in effect at the time of construction, and the following. Required emergency escape and rescue openings shall be operational from the inside of the room without the use of keys or tools. Bars, grilles, grates or similar devices are permitted to be placed over emergency escape and rescue openings provided the minimum net clear opening size complies with the code that was in effect at the time of construction and such devices shall be releasable or removable from the inside without the use of a key, tool or force greater than that which is required for normal operation of the escape and rescue opening.

[F] SECTION 703
FIRE-RESISTANCE RATINGS

703.1 Fire-resistance-rated assemblies. The required fire-resistance rating of fire-resistance-rated walls, fire stops, shaft enclosures, partitions and floors shall be maintained.

703.2 Opening protectives. Required opening protectives shall be maintained in an operative condition. All fire and smokestop doors shall be maintained in operable condition.

Fire doors and smoke barrier doors shall not be blocked or obstructed or otherwise made inoperable.

[F] SECTION 704
FIRE PROTECTION SYSTEMS

704.1 General. All systems, devices and equipment to detect a fire, actuate an alarm, or suppress or control a fire or any combination thereof shall be maintained in an operable condition at all times in accordance with the *International Fire Code.*

704.2 Smoke alarms. Single or multiple-station smoke alarms shall be installed and maintained in Groups R-2, R-3, R-4 and in dwellings not regulated in Group R occupancies, regardless of occupant load at all of the following locations:

1. On the ceiling or wall outside of each separate sleeping area in the immediate vicinity of bedrooms.

2. In each room used for sleeping purposes.

3. In each story within a dwelling unit, including basements and cellars but not including crawl spaces and uninhabitable attics. In dwellings or dwelling units with split levels and without an intervening door between the adjacent levels, a smoke alarm installed on the upper level shall suffice for the adjacent lower level provided that the lower level is less than one full story below the upper level.

Single or multiple-station smoke alarms shall be installed in other groups in accordance with the *International Fire Code.*

704.3 Power source. In Group R occupancies and in dwellings not regulated as Group R occupancies, single-station smoke alarms shall receive their primary power from the building wiring provided that such wiring is served from a commercial source and shall be equipped with a battery backup. Smoke alarms shall emit a signal when the batteries are low. Wiring shall be permanent and without a disconnecting switch other than as required for overcurrent protection.

> **Exception:** Smoke alarms are permitted to be solely battery operated in buildings where no construction is taking place, buildings that are not served from a commercial power source and in existing areas of buildings undergoing alterations or repairs that do not result in the removal of interior wall or ceiling finishes exposing the structure, unless there is an attic, crawl space or basement available which could provide access for building wiring without the removal of interior finishes.

704.4 Interconnection. Where more than one smoke alarm is required to be installed within an individual dwelling unit in Group R-2, R-3, R-4 and in dwellings not regulated as Group R occupancies, the smoke alarms shall be interconnected in such a manner that the activation of one alarm will activate all of the alarms in the individual unit. The alarm shall be clearly audible in all bedrooms over background noise levels with all intervening doors closed.

Exceptions:

1. Interconnection is not required in buildings which are not undergoing alterations, repairs, or construction of any kind.

2. Smoke alarms in existing areas are not required to be interconnected where alterations or repairs do not result in the removal of interior wall or ceiling finishes exposing the structure, unless there is an attic, crawl space or basement available which could provide access for interconnection without the removal of interior finishes.

CHAPTER 8

REFERENCED STANDARDS

This chapter lists the standards that are referenced in various sections of this document. The standards are listed herein by the promulgating agency of the standard, the standard identification, the effective date and title and the section or sections of this document that reference the standard. The application of the referenced standards shall be as specified in Section 102.7.

ASME

American Society of Mechanical Engineers
Three Park Avenue
New York, NY 10016-5990

Standard reference number	Title	Referenced in code section number
A17.1—2004	Safety Code for Elevators and Escalators with A17.1a — 2005 Addenda and A17.15 Supplement 2005 . . . . 606.1	

ASTM

ASTM International
100 Barr Harbor Drive
West Conshohocken, PA 19428-2959

Standard reference number	Title	Referenced in code section number
F1346—91 (2003)	Performance Specifications for Safety Covers and Labeling Requirements for All Covers for Swimming Pools, Spas and Hot Tubs 303.2	

ICC

International Code Council
500 New Jersey Avenue, NW
6th Floor
Washington, D.C. 20001

Standard reference number	Title	Referenced in code section number
ICC EC—06	ICC Electrical Code® — Administrative Provisions 201.3, 604.2	
IBC—06	International Building Code® 102.3, 201.3, 401.3, 702.3	
IFC—06	International Fire Code® 201.3, 702.1, 702.2, 704.1, 704.2	
IFGC—06	International Fuel Gas Code® 102.3	
IMC—06	International Mechanical Code® 102.3, 201.3	
IPC—06	International Plumbing Code® 201.3, 505.1, 602.2, 602.3	
IZC—06	International Zoning Code® 102.3, 201.3	

INDEX

2006 INTERNATIONAL PROPERTY MAINTENANCE CODE®

U

UNOBSTRUCTED
Access to public way. 702.1
General, egress 702.1

UNSAFE
Equipment 108.1.2
Existing remedies 102.4
General, condemnation 108, 110
General, demolition 110
Notices and orders 107, 108.3
Structure 108.1.1

USE
Application of other codes 102.3
General, demolition 110

V

VACANT
Closing of vacant structures 108.2
Emergency measure .109
Method of service 107.3, 108.3
Notice to owner or to person
responsible 107, 108.3
Placarding of structure 108.4
Vacant structures and land. 301.3

VAPOR
Exhaust vents. 302.6
Process ventilation .403.4

VEHICLES
Inoperative 302.8
Painting. 302.8

VENT
Connections 504.3
Exhaust vents. 302.6
Flue. 603.2

VENTILATION
Clothes dryer exhaust. 403.5
Combustion air 603.5
Definition 202
General, ventilation .403
Habitable rooms 403.1
Process ventilation .403.4
Recirculation 403.2, 403.4
Toilet rooms 403.2

VERMIN
Condemnation 108
Insect and rat control 302.5, 308

VIOLATION
Condemnation 108
General 106
Notice 107, 108.3
Penalty 106.4
Placarding of structure 108.4
Prosecution. 106.3
Strict liability offense. 106.3, 202
Transfer of ownership 107.5

W

WALK
Sidewalks 302.3

WALL
Accessory structures 302.7
Exterior surfaces 304.2, 304.6
Exterior walls 304.6
Foundation walls. .304.5
General, fire-resistance rating 703.1
Interior surfaces 305.3
Outlets required 605.2
Temperature measurement 602.5

WASTE
Disposal of garbage 307.3
Disposal of rubbish 307.2
Dwelling units 502.1
Garbage storage facilities 307.3.1

WATER
Basement hatchways 304.16
Connections 506.1
Contamination. 505.2
General, sewage 506
General, storm drainage 507
General, water system. 505
Heating 505.4
Hotels 502.3
Kitchen sink 502.1
Required facilities 502
Rooming houses. .502.2
Supply. 505.3
System. 505
Toilet rooms 503
Water heating facilities 505.4

WEATHER, CLIMATE
Heating facilities. 602
Rule-making authority. 104.2

WEATHERSTRIP
Window and door frames 304.13

WEEDS
Noxious weeds 302.4

WIDTH
Minimum room width. 404.2

WIND
Weather tight. 304.13
Window and door frames 304.13

WINDOW
Emergency escape. 702.4
Glazing. 304.13.1
Guards for basement windows. 304.17
Habitable rooms 402.1
Insect screens. 304.14
Interior surface 305.3
Light. 402
Openable windows 304.13.2
Toilet rooms 403.2
Ventilation 403
Weather tight. 304.13
Window and door frames 304.13

WORKER
    Employee facilities 503.3, 602.4
WORKMANSHIP
    General 102.5

EDITORIAL CHANGES – THIRD PRINTING

Page 21, ASME line 1 now reads . . . A17.1–2004 . . . — 2005 Addenda and A17.15 Supplement 2005

INTERNATIONAL CODE COUNCIL®

People Helping People Build a Safer World™

Membership Application
This form may be photocopied

MEMBERSHIP CATEGORIES AND DUES*— ANNUAL MEMBERSHIP
Special membership structures are also available for Educational Institutions and Federal Agencies.
For more information, please visit www.iccsafe.org/membership or call 1-888-ICC-SAFE (422-7233), x33804.

GOVERNMENTAL MEMBER**
Government/Municipality (including agencies, departments or units) engaged in administration, formulation or enforcement of laws, regulations or ordinances relating to public health, safety and welfare. Annual member dues (by population) are shown below. Please verify the current ICC membership status of your employer prior to applying.

☐ Up to 50,000.........$100   ☐ 50,001–150,000........ $180   ☐ 150,001+.......... $280

**A Governmental Member may designate 4 to12 voting representatives (based on population) who are employees or officials of that governmental member and are actively engaged on a full- or part-time basis in the administration, formulation or enforcement of laws, regulations or ordinances relating to public health, safety and welfare. Number of representatives is based on population. All dues for representatives have been included in the annual member dues payment. Please call 1-888-ICC-SAFE (422-7233), x33804 for information about how to designate your voting representatives.

☐ **CORPORATE MEMBERS ($300)**   An association, society, testing laboratory, manufacturer, company or corporation

INDIVIDUAL MEMBERS
☐ **PROFESSIONAL ($150)**   A design professional duly licensed or registered by any state or other recognized governmental agency

☐ **COOPERATING ($150)**   An individual who is interested in International Code Council purposes and objectives and would like to take advantage of membership benefits

☐ **CERTIFIED ($75)**   An individual who holds a current Legacy or International Code Council certification

☐ **ASSOCIATE ($35)**   An employee of a current ICC Governmental Member

☐ **STUDENT ($25)**   An individual who is enrolled in classes or a course of study including at least 12 hours of classroom instruction per week

☐ **RETIRED ($20)**   A former governmental representative, corporate or individual member who has retired

New Governmental and Corporate Members will receive a free package of 7 code books. New Individual Members will receive one free code book. Upon receipt of your completed application and payment, you will be contacted by an ICC Member Services Representative regarding your free code package or code book. For more information, please visit www.iccsafe.org/membership or call 1-888-ICC-SAFE (422-7233), x33804.

Please print clearly or type information below:

Name
_____

Name of Jurisdiction, Association, Institute or Company, etc.
_____

Title
_____

Billing Address
_____

City _____ State _____ Zip+4 _____

Street Address for Shipping
_____

City _____ State _____ Zip+4 _____

E-mail
_____

Telephone
_____

Tax Exempt Number (If applicable, must attach copy of tax exempt license if claiming an exemption)
_____

Payment Information:

VISA, MC, AMEX or DISCOVER Account Number   **Exp. Date**

Return this application to:
International Code Council
Attn: Membership
5360 Workman Mill Road
Whittier, CA 90601-2298

Toll Free: 1-888-ICC-SAFE (1-888-422-7233), x33804
FAX: (562) 692-6031 (Los Angeles District Office)
Or, apply online at **www.iccsafe.org/membership**.
Please refer to Tracking Number 66-05-274 when applying.

If you have any questions about membership in the International Code Council, call 1-888-ICC-SAFE (1-888-422-7233), x33804 and request a Member Services Representative.

REF 66-05-274

*Membership categories and dues subject to change.
Please visit www.iccsafe.org/membership for the most current information.